Einstein's Ignorance of Dark Energy

"From physics evolution...
...to view point revolution"

This book describes the discovery of the origin and the nature of Dark Energy and Dark Matter that Albert Einstein ignored in his descriptions of the general theory of relativity because of not considering processes below event horizons of length and time. It shows that masses are detached energies in the wrong space-time quadrants, banished, trying to make their way home via black holes.

Einstein missed this extension of four-dimensional space-time in his general theory of relativity because he did not consider the basic set-up of space-time alone as a vacuum energy construction with an overall distortion pattern, caused by four interacting energy quadrants that converted quantized energy portions into masses by joint forces.

An extended and diversified space-time reveals energy quadrants below event horizons of length and time, causing space inflation. A reversed vacuum energy tank appears as dark matter halos, holding the galaxies together. This new model confirms precisely the experimentally observed ratio between baryonic mass, dark energy, and dark matter of the Wilkinson Microwave Anisotropy Probe (WMAP).7-year result, published in October 2010, confirming 4.6% baryonic matter, 72.24% dark energy and 23.16% dark matter. It combines peculiarities of quantum mechanics with Einstein's general theory of relativity.

The gravitational constant "G" is derived in a revolutionary way. Trigonometry expresses precisely the special theory of relativity, but substantiates well-known, expected and not yet discovered peculiarities of quantum mechanics. The model explains why evaluations of light interferences differ, depending on the observation of light as a particle stream, or wave. It shows that there is no big mystery about the particle-wave dualism of physics.

ISBN 978-1-4475-2467-0

© 2011, Dr. Henryk Frystacki
henryk.frystacki@t-online.de
Edition April 2011

Contents

I) Introduction — 5

II) Revolutionary new views on Einstein's space-time — 7

III) Discussion of space-time leaps and conclusions — 14

IV) Definitions for the construction of a rotary time suite — 16

V) Gravitational time dilation within the super symmetry — 33

VI) New derivation of gravitational constant G and masses — 35

VII) Expedition through the standard model of physics — 37

VIII) References — 45

I) Introduction

The special theory of relativity[1] "STR" and the general theory of relativity[2] "GTR" introduce speed of light scaled time as fourth dimension, in addition to our three perceived space dimensions length, width, height. Euclidean geometry has been replaced by differential geometry. The GTR turns into the STR in case of sufficiently small areas in space-time, or an ideal, mass-free universe.

STR and GTR are based on the assumption and on many experimental confirmations of the constancy of speed of light in a vacuum[3], observed independently of any own motion or any other type of energetic influence. The consequences of this constancy of speed of light are time dilations, length contractions and a relativity of simultaneity of events. STR describes additionally an energy equivalence of baryonic masses with $E=mc^2$. GTR explains gravity by curvatures of space-time. The constancy of the speed of light in a vacuum was initially a postulate, stemmed from the assumption that the speed of light in Maxwell's equations of electromagnetism[4] stays constant in any inertial frame of reference, assuming homogeneous time and also homogeneous and isotropic space. Minkowski replaced Einstein's original concept of separated space and time by the construction of "space-time" and reformulated Einstein's work[5]. The paths of light in Minkowski's space-time-graph cover mathematically a zero space-time interval, having the impact that any observer will read the same constant value for the speed of light.

Asking Einstein about the nature of time, he answered: "time is what you read on a clock", expressing that time describes the sequence of events, moving on in the present, with an origin in the past and heading towards the future. The present of an observation of any event can be only defined in one single point in space-time. Other points neither placed in the past nor in the future are separated in space. Keeping the speed of light constant, space-time can be described with curvatures, as done by Einstein with energy tensors and relativistic field equations, explaining a far-reaching gravitation despite the restriction for information and energy transport at the speed of light. Einstein's cosmological constant Λ[6], introduced and later dismissed by him, was re-introduced to adapt the general theory of relativity to the latest discoveries and current models of astrophysics[7]. However, STR and GTR have no proper explanation for the uncertainty principle of Heisenberg[8], stating that certain pairs of physical properties like position and momentum cannot be known to arbitrary precision. At known types of energy quantum levels, both are inconsistent with quantum mechanics. Two equations of relativistic mechanics describe time dilation and length contraction of the special theory of relativity STR with the coordinates x', y', z' and t' of a moving system S' at the speed v and with the coordinates x, y, z and t of the initial starting system S. Δx and $\Delta x'$ are lengths, Δt and $\Delta t'$ are time intervals read on clocks and compared with each other. System S' moves along the x-axis. c stands for the speed of light in a vacuum.

$$x = \frac{x'+vt}{\sqrt{1-v^2/c^2}} \qquad t = \frac{(t'+\frac{v}{c^2}x')}{\sqrt{1-v^2/c^2}}$$

Note: Time in non-coinciding x'-coordinates in S' differs, if evaluated in S. This describes a relativity of simultaneity of events. For two events in the same space location in S' ($x'_2 = x'_1$) but at a different time $t'_2 \neq t'_1$, time dilation of the special theory of relativity is described with $x'_2 - x'_1 = 0$ by:

$$\Delta t = \frac{\Delta t'}{\sqrt{1-v^2/c^2}} + \frac{v}{c^2} \frac{x'_2-x'_1}{\sqrt{1-v^2/c^2}} = \frac{\Delta t'}{\sqrt{1-v^2/c^2}} \geq \Delta t'$$

The time period of a moving observer is shorter in comparison with the time period of any remaining observer in the starting point: A moving observer stays continuously in the present of the remaining one but ages slower. The corresponding length contraction is based on proven constancy of speed of light, with the impact that the original distance Δx to a destination is getting shorter for the moving observer, expressed by $\Delta x'$:

$$\Delta x = \frac{\Delta x'}{\sqrt{1-v^2/c^2}} \geq \Delta x'$$

Limiting time and length above an individual observer's event horizon with a Planck-time and a Planck-length and transforming all translations in three-dimensional space in an avant-garde way into rotary processes in space-time, results in complementary views on time, length, speed and space-time curvatures of the general theory of relativity GTR: Quantum mechanics gets its feasible "space-time view" on Heisenberg's uncertainty principle. A rotary set-up of space-time reveals interchangeable space-time dimensions and an overall super symmetry of space-time. It allows the discussion of the special theory of relativity and the general theory of relativity in combination with all quantum mechanical aspects, mass generation and dark energy sources of an expanding universe.

II) Revolutionary new views on Einstein's space-time

Cosmology assumes dark energy to explain the expansion of space with increasing speed, and dark matter to understand the stability of galaxies. The standard model of cosmology, the Lambda-CDM model, includes the cosmological constant Λ-Lambda in Einstein's formulas of general relativity to consider the expansion of space. The observations of distance-red-shift relations of supernovae gave new significance to this constant[9, 10, 11, 12]. The following discussion of newly diversified space-time energy describes three space-time quadrants below the event horizons of length and time that complete the present picture of space-time. One quadrant of these three energy reservoirs appears with static energy, the second one with reversed energy and the third with dynamic energy. All four quadrants expand space with a roll-out across the entire set-up. The reversed energy field tank below the event horizon of length and of time causes additionally gravity in the length-time-grid of space-time, holding the galaxies together. This roll-out process adjusts the cosmological constant and could be a natural way to derive the cosmological constant that is used in cosmology from quantum physics. Measurements have shown that it is in the order of about 10^{-120} reduced Planck units which is by the factor 10^{-120} smaller than assumed by quantum physics on Planck scale, with the completely wrong assumption of quantum mechanics physicists that quantum vacuum is equivalent to the cosmological constant[13,14]. This assumption turned out to be the greatest error ever in physics that has been scientifically postulated and published.

The Planck length is related to Planck energy by the uncertainty principle. At Planck scale of quantum mechanics the concept of space-time collapses as quantum indeterminacy becomes absolute. The Compton wavelength gets already into the range of the Schwarzschild radius of a black hole. The framework of the quantum field theory describes fundamental interactions and forces on quantum levels but cannot yet integrate gravity in a way that all fundamental forces are unified at this Planck scale. String theories, loop quantum gravity and non-commutative geometry are attempts to integrate gravity into the quantum field theory. The discussion of the diversified energies of an extended space-time model does not support itself on any of these gravity quantization theories but focuses on energy transformation aspects of space-time across event horizons that are not yet covered by any current model of quantum mechanics.

A diversified energy reservoir below the event horizon for length and time and its interactions can be already studied in a reduced two-dimensional space-time grid. A reduced two-dimensional space-time grid of the special theory of relativity uses an x-axis for length and a y-axis for time. The general theory of relativity introduces mass caused curvatures into this picture that distort this two-dimensional grid into one direction of a third z-dimension. Staying in this grid of finished length and passed time above event horizons with x and y leads to the well-known formulas of Planck length and Planck time, using the reduced Planck constant, the gravitational constant and the speed of light.

The extension of this two-dimensional grid of positive units for finished lengths and for passed time with opposing negative values that mark energy density levels below the event horizons of length and time leads to a space-time grid with four space-time quadrants instead of one space-time quadrant with length, time and vacuum energy density. Staying below a Planck length and the Planck time makes these negative values possible without any inversion of time or turning space inside out because they affect only the vacuum energy density, but now within a new picture with three different energy tanks. Entering another space-time quadrant, it develops its full dimensions and the original length-time quadrant disappears below new event horizons.

The chosen geometry for these four quadrants is an extension of the space-time diagrams of Robert W. Brehme[15,16,17]. These diagrams have been published in the American Journal of Physics and have been only used for educational purposes. An extended view with four quadrants of space-time brings scientific aspects into this teaching method: Three additional space-time quadrants below and at the event horizons of length and time deliver 75% of the energy contributions for the space expansion by interaction with the first space-time quadrant and baryonic masses. The (-x, -y)-space-time quadrant opposes the (+x, +y)-quadrant which generates a 25% energy effect of cold dark matter by the grid distortions. A condensation of about 4.6% (to within 0.1%) of baryonic matter and their kinetic energies reduce the share of dark energy from 75% to about 72.1% (to within 1.5%) and of dark matter from 25% to about 23.3% (to within 1.3%) if these shares are taken from current estimates of cosmology and the WMAP seven years analysis[18,19]. A model of three space-time quadrants of dark energy below event horizons of length and time will represent these shares accurately.

The discussion does not focus on the exotic elementary particle zoo that may condense to hot dark matter by locally limited interaction of the four space-time quadrants. Any condensed particle affects an extended space-time grid in a corresponding way to baryonic masses.

Length contraction, time dilation and the relativity of simultaneity of events of the special theory of relativity can be derived from the diagram in figure 1. A similar type of diagram has been introduced by Robert W. Brehme. In contrary to Minkowski's well-known two-dimensional space-time diagram, Brehme's diagram maintains the linear unchanged scales for all observers as it actually happens. Figure 1 combines a time dilation diagram on the upper left side with and a length contraction diagram on the upper right side by the use of a common y-axis. The relativity of the simultaneity of events makes this coincidence of a length axis of an inertial frame of reference at the speed of light and the time axis of a resting observer feasible, as the further discussion will show.

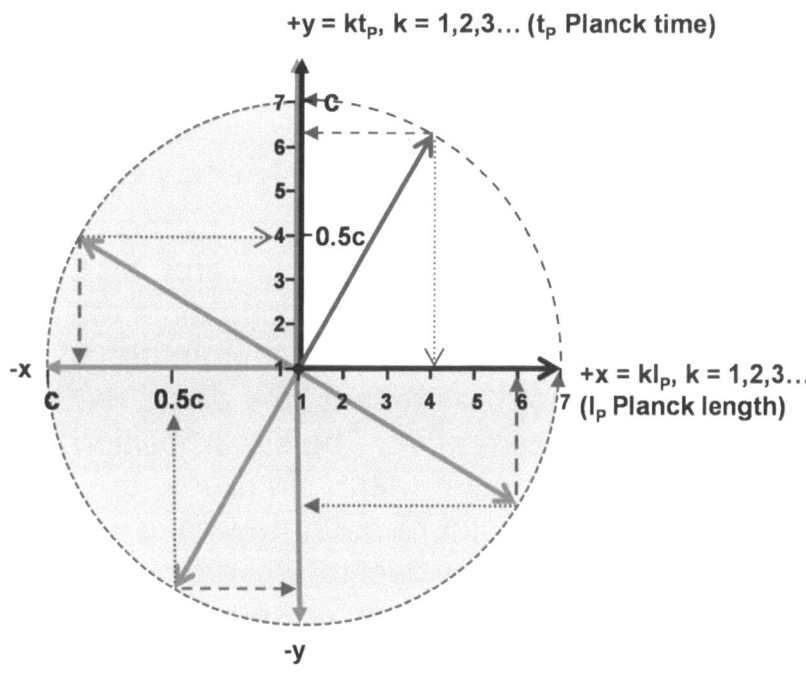

Figure 1: Length contraction and time dilation with four space-time quadrants

The (+x)-axis describes a length of an intended motion direction within a mass-free inertial frame of reference of a resting observer, subdivided into Planck length distances. The (+y)-axis shows the time on a clock of this resting observer, running at the same steady pace in each position on the x-axis. The right side of figure 1 describes length contraction by projections of rotating lengths onto the x-axis and the caused change of simultaneity of events by the projection of this rotating length onto the y-axis. A contraction will be read differently by the resting observer or by the moving observer: The resting observer notices a reduction of the size of the moving inertial system and the moving observer notices a reduction of the original space distance to the target. The rotation angle has to be matched with the relative speed increase of the moving observer. For this purpose any freely chosen length is rotated from the x-axis into the y-axis and then calibrated with a linear superimposed speed scale from 0 up to the speed of light, as demonstrated in figure 1. The Pythagoras formula leads to correct length contraction values as x-projections on the x-axis and to the correct readings of the change of simultaneous events on any rotated x-length into serial events for the resting observer by projections of the rotating length onto the y-time axis.

Using the Pythagoras formula $l^2(\frac{v^2}{c^2}) + x^2 = l^2$ results in length contraction: $x = l\sqrt{1 - v^2/c^2}$. The length l is any unaltered length in the moving system, $l\frac{v}{c}$ the projection value on y if we rotate l fully into y and stay with its length calibration. x reflects the reading of the length l by the resting observer on x by comparison with the unchanged x-scale and at the same time the reading of the moving observer of the shrinking distance towards the target in space. Note that we have neither shifted coordinates of the moving object along x due to the speed, nor the present time of the resting observer along the y-axis, indicated by the zero point: The superposition of two diagrams in the starting origin of a measurement simplifies the discussion of the energetic impacts. The zero point of each new measurement leaps for the resting observer along the y-axis due to the progress of time if the initial frame of reference of the resting observer serves as a frozen reference. A monitored length of the moving frame of reference leaps additionally Planck length by Planck length along the x-axis. However, this is not necessarily increasing the distance to the resting observer as it is possible to change the heading after each Planck time into any direction, even into an orbit around the resting observer. In case of translation along x of a Cartesian system, the zero-x-y-coordinates of the moving frame of reference shift in the view of the resting observer between the y-axis (zero Planck length per Planck time) and the bisector of x- and y-axis (speed of light with one Planck length per Planck time).

The Pythagoras formula can be applied in an identical way for the left side of figure 1. This way, the projection of an original time interval on the y-axis that rotates now from y to (-x) shows the reading of a decreasing time interval for a moving observer in relation to the resting observer: The time of the moving observer has been dilated in relation to the resting observer, causing the well-known fact that the moving observer reads a shorter time span between events in the original environment than the resting observer does. The relatively stretched time scale of the moving observer is merely characterized by the reciprocal factor: $1/\sqrt{1-v^2/c^2}$. Speed stays constant for both observers within their inertial frame of reference because length and time change by the same ratio. Speed vectors can be easily added in the rotational construction across several inertial frames of references at different relative speed. The angle between the own time axis and the own length axis of any observer maintains always 90°. The energy tanks below event horizons for length and time make this possible. The (-x)-axis reflects the strength of time dilation that accommodates all events of any original time interval below the event horizon of time, i.e. without progressing y-time of the resting observer. The change of simultaneity for a resting observer by observation and rotation of a constant length l from the x-axis to the y-axis can be described by the formula part of relativistic mechanics for time intervals in moving systems[20]:

$$\Delta t = \frac{v}{c^2} \frac{l}{\sqrt{1-v^2/c^2}}$$

The term $1/\sqrt{1-v^2/c^2}$ compensates the length contraction $l = x\sqrt{1-v^2/c^2}$ that a resting observer will monitor, in other words it keeps the value of the contracting length *l* constant and equal to the initial x-value. One speed of light factor *c* changes this length value into a corresponding time difference, or, more precisely, into a corresponding time delay Δt of an event. The remaining quotient *v/c* reflects the projection value onto the y-axis. This projection is linked to the rotation angle that is ruled by this quotient *v/c*.

This rotational picture is completed by the axis (-y) that opposes the y-axis in the same way that (-x) is opposing x. This (-y)-axis can be understood as a clockwise rotated axis in relation to the x-axis, or anticlockwise rotated axis in relation to the (-x)-axis. Its value stays below a Planck time in relation to the y-axis, just like (-x) stays below a Planck length in relation to the x-axis. The (-y)-axis reflects the strength of a length contraction that accommodates all events on any original length below the event horizon, i.e. actually not being any length with noticeable events for the resting observer. (-x) and (-y) are neither a length nor a time in the environment of any observer with x-length and y-time but summarize all energy processes of the space-time grid below the event horizons for length and time on base of action and reaction and rolled up or out via the rotational construction. A (-y)-axis shows maximum accelerated processes below the event horizon for any observer with an existing x-length in the same way that x shows maximum acceleration of processes below the event horizon for any observer with an inertial frame of reference at the speed of light in relation to its origin with x-length and y-time. The rotational construction proves that space-time frames of any size and without any rest mass can only reach a maximum of relative speed of light.

The three space-time quadrants below the event horizon of length and time deliver with their energy tanks 75% of the energy contributions for the roll out of space on top of all interactions of the (+x, +y)-quadrant with the baryonic masses. The (-x, -y)-space-time quadrant opposes this (+x, +y)-quadrant and generates a 25%-energy effect of cold dark matter by the grid distortions. A condensation of about 5% of baryonic matter energy and their kinetic energies reduce this share of dark energy from 75% to about 72% and of dark matter from 25% to about 23%. This could be explained by an energy origin of baryonic matter of about 1% contributed by the (-x, -y)-quadrant, 1% by the (-x, +y)-quadrant, 1% by the (+x, -y)-quadrant and 1% by the (+x, +y)-quadrant. These precipitated baryonic masses are entirely attributed to the (+x, +y)-length-time-quadrant. A seeming fifth percentage point of baryonic matter can be explained by a 2% reduction of the dark matter share if the (+x, +y)-share and the (-x, -y)-share of precipitated masses are both considered in the space-time curvatures of the (+x, +y)-quadrant that are caused by the presence of these masses. The mentioned rounded off mass share corresponds to the current WMAP estimates[19]. A value of 4.6% for masses leads to 72.24% dark energy and 23.16% dark matter. The latest WMAP-results show the difference of 0.14% shifted from dark energy to dark matter. The total rest mass rotates summarizing length-time axes of coordinates anticlockwise in figure 1 like kinetic energy and any other form of energy, according to the principle of equivalence. Four-dimensional space-time requires causally related distribution. Einstein's general theory of relativity captures only this anticlockwise rotation together with the local contractions by Ricci-tensors and relativistic field equations. It does not consider the basic set-up of space-time alone as a vacuum energy construction with an overall distortion pattern that is caused by four space-time quadrants. Figure 1 demonstrates that space-time alone, i.e. without any precipitation of masses, is already rolling out space and embeds the precipitated galaxies of masses with a dark cold matter effect of the (-x, -y)-quadrant.

All four quadrants expand space with a roll-out across the four axes. As opposing axes are completely below the event horizon of each other and successive axes of the rotational construction exactly at the event horizon of each other with quantized release of energy, the whole potential energy of each axis and the opposition of quadrants have to be considered for a balanced roll out of length and time. Each roll out of a Planck length affects the whole construction, adjusting the cosmological constant far below the value of models where quantum vacuum is equivalent to the cosmological constant. The measured cosmological constant is by factor of 10^{-120} smaller, indicating distributed energies and slowdown of space inflation by a rollout across their space-time quadrants.

Length appearance, length contraction, time appearance and time dilation can be connected with energy tanks below the event horizons of length and time. These energy tanks seem to cause dark energy and dark matter of the current models of cosmology. Planck length and Planck time are the observations of the finished process above the event horizon. They form a quantum grid on the base of Planck energy and Heisenberg's uncertainty principle. Planck length l_P and Planck energy E_P are linked with the formula $l_P = E_P G/c^4$, inserting $E_P = 1.9561 \cdot 10^9$ [J] $= 1.2209 \cdot 10^{28}$ [eV], applying the gravitational constant G and considering a reduction by c^4 of superimposed c-calibrations of the four perpendicular axes. Scaling time with factor c like in the general theory of relativity leads to equal values of l_P and t_P because of c=1 and $t_P = E_P G/c^5$. The values of l_P and t_P are additionally equal to E_P if G is scaled up to 1. The smallest quantized values of (-x) and (-y) contain both E_P, acting against the energies of x and y by the interaction principle of kinetic energy with static potential energy and across the rotary set-up. The (+x, +y)-field reacts to the presence of masses with (-x, -y)-impacts, as captured by the general theory of relativity: Einstein's curvatures of space-time describe changes of these space-time energy densities.

The rotational construction of an extended space-time makes it possible to talk about an energy grid that accommodates the effects of the general theory of relativity and quantum mechanics, but also the phenomena of dark energy, dark matter and baryonic matter. The mass of baryonic matter can be interpreted as a quantized interaction with the (-x, -y)-tank, making the (-x, -y)-area a candidate for a mass background field[21,22]. Masses are based in this model on an interaction of partially rotated space-time zones with reverse tanks, being subject to major grid distortions. The static sector delivers the electric charge of an elementary particle or nucleus formation, the dynamic sector the magnetic spin. The time dilation has been covered by the special and general theories of relativity as only one function of the (+y,-x)-tank. The triple split into static (+y,-x)-tank, reverse (-x,-y)-tank and dynamic (+x, -y)-tank are new aspects for the discussions about space, time, matter and vacuum energy density on Planck scale. The cosmological constant reflects the constant negative pressure of the adiabatic roll out of space acting as repulsive force to gravity.

III) Discussion of space-time leaps and conclusions

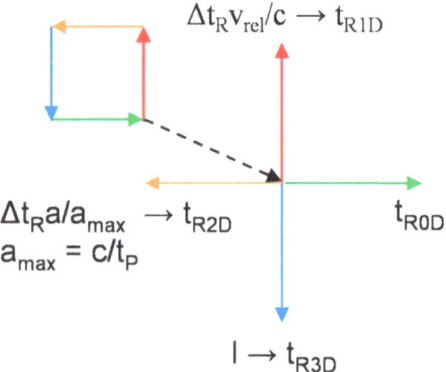

Figure 2: Space-time-speed-acceleration leap frame

The frame of figure 2 that is related to figure 1 with $x = t_{R3D}$, $y = t_{R0D}$, $-x = t_{R1D}$ and $-y = t_{R2D}$ is expected to have the following features:

- The rotational processes limit the speed of mass-less frames at speed of light. Matter acquires a kind of passive speed of light on $-y = t_{R2D}$ by its generation: The production of elementary particles out of photons, which is state of the art, tells that these particles should have speed of light against an appropriate reference point in space-time. Baryonic masses, however, reach relative speed of light after their completion process relatively to their environment only by a mass defect with an EMP of this released space-time energy, dispersing at speed of light.

- Any relative rotation of elementary frames by 180° is supposed to generate stable negative acceleration densities with the appearance of matter. Figure 3 explains this process, keeping original parameters and quantities of a rotated cut out fragment despite of such a rotation.

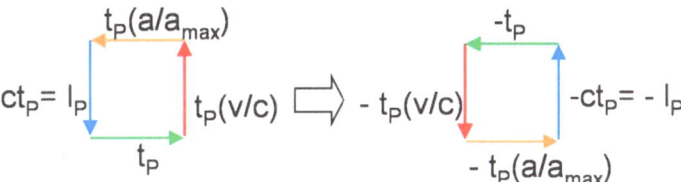

Figure 3: 180°- turn of space-time-speed-acceleration components

Figure 3 shows the principle of the generation of matter for an observer with t_{R0D}-time: 180° rotated frames appear with negative acceleration towards each other. The frames are generated in pair as (180°)-matter detachment and (-180°)-antimatter detachment by appropriate physical cause and effect scenarios. The confrontation of ±180°-rotated units with their space-time environment causes the contraction of the surrounding space and its time dilation. Spatial aggregations of frames can accumulate negative quantities, increasing the strength of gravity towards each other and the time dilation in the surrounding environment. The particles are still embedded in the overall (-x, -y)-quadrant of space-time if they are attributed to (+x, +y)-quadrant.

The system immanent quasi passive speed of light on $-y = l_{R2D}$ does not conflict with the fact that baryonic masses cannot reach speed of light on l_{R3D}: The fundamental problem of any 180°-rotated unit of not being able to reach relative speed of light can be derived from figure 3: The acceleration component replaced the Planck time component and any relative speed is now dilating this acceleration component with relative energy storage. This negative acceleration component reflects a precipitated mass of the rotated unit. Therefore, any mass increases its value with the relativistic factor. All events have to be unambiguously assigned by cause and effect in order to correctly calculate relative developments and changes by accumulated frames for an observer with defined location and motion in space-time. The static electrical charge and dynamic magnetic momentum can be explained by acquired angular leap frame momentum during the particle generation by rotation and precipitation and on base of active and passive speed of light aspects. Rotation of two ±90° frames by ±180° is supposed to produce electrically and magnetically neutral pairs of elementary particles, because of electrically and magnetically neutral angular frame momentum in space.

IV) Definitions for the construction of a rotary time suite

- Four-dimensional space-time is based on distributed energy across four interacting space-time quadrants that produce an observer's individual space and time, depending only on the location within the four quadrants.

- Inertial space-time energy frames at passive speed of light in relation to appropriate reference points in the overall space-time set-up results in manifestations of space lengths and simultaneity of events along these space lengths.

- Relative speed of light of baryonic masses in an extended space-time construction with four axes, measured against suitable reference points, is fully in line with the standard model of physics as elementary particles can be produced with photons that moved at the speed of light in relation to the pair of matter particles they end up.

- The electrical charge of free particles or nuclear particle aggregation is a residual quantized component of the static space-time energy quadrant and the magnetic spin of this particle or nuclear particle is the residual component of the dynamic space-time energy quadrant. Both have been detached during the generation process of particle and atomic nucleus.

- Active speed of light of inertial space-time energy frames generates for a resting observer sequential events along the inertial frames of reference at this active speed of light. This corresponds to quantum mechanics.

- Any kind of energetic processes is describable either by relative rotations of space-time coordinates in an extended space-time reference diagram, or by relative expansion or contraction of coordinates on the unchanged location axes of such diagrams. These four axes are scaffolding of the rotational energy roll out in clockwise direction, or anticlockwise roll up.

- Symmetrical energy processes appear with an asymmetry in case they are not evaluated from a central position in space-time, for example from one of the four scaffolding axis of space-time.

- Space-time flow and tension scenarios of a rotational symmetry can be simplified by trigonometric projections of proportions between an object's time t_R and a basic system time t_{0D}, using the reference time interval t_{0F}

and unchanged calibrations of rotating time and length. This includes the influence of masses according to the equivalence principle of energies.

- The anticlockwise rotations and contractions caused by the presence of masses on base of cause and effect and upon the existing environment of an observer across four-dimensional space-time lead to Ricci-tensors and the relativistic field equations of the general theory of relativity. This approach catches the influence of baryonic masses and other detectable energies. It does not consider a basic embedding in four different space-time quadrants with their own interacting vacuum energies.

- The expansion of space at an increasing speed is taken into account by the cosmological constant, but without a clear understanding of its origin. The rotational set-up and the opposing axes define the inflation of space.

- A trigonometric approach in a two-dimensional length-time grid simplifies considerably the discussion of the principles, if it uses the reference of an observer's frozen starting grid of rotating length and time coordinates. The extension to Einstein's four-dimensional space-time is realized by a simple causal distribution of the processes in three-dimensional space. It does not affect the validity of any finding for the two-dimensional grid.

- A time dilation of the special theory of relativity SRT can be geometrically explained with a graph, showing two perpendicular time axes t_{0D} and t_{1D} and the turning object's time line t_R between those two. Such diagrams have been already introduced to the scientific community for pedagogic purposes. The advantages of such geometry are unchanged scales and values of time and length in any observer's own environment. This is fully corresponding to all experimental evaluations in inertial systems.

- Any straight timeline t_R shows a nonlinearly increasing angle against the straight timeline t_{0D} if it is rotated around the origin of time measurement in a frozen reference frame by the linear increase of relative speed of the object with time t_R in comparison to its inertial starting frame of reference with time t_{0D}.

- The end of a t_R-reference interval moves on a quarter-circle between t_{0D} and t_{1D} because of the constant speed of light and constant structural dimensions of the moving object. Therefore, it is possible to capture the rotations by trigonometric functions and a reference time interval t_F.

- The t_R-dilation works in mutual directions between inertial systems, but depending on the relative acceleration history of these systems on base of cause and effect. The acceleration component of a space-time-speed-acceleration frame clarifies this direction of dilation, unambiguously.

- Not knowing the origin of a relative motion and its history leads to the uncertainty of impulse power versus the exact location in space and time around the individual environment, because of not being in a position to identify whether simultaneous events will change into serial events and the shortage of distances, or into simultaneous events and an increase of distances. The same problem arises in fixed space-time positions without a possibility to capture energetic percussions on this location, being thus the effects and not the causes. Only the time-length-speed-acceleration frame can achieve the rotational effect in both directions of two observed objects, keeping all perceived dimensions and the pace of time on each of both inertial frames of reference constant, including the constancy of the speed of light.

- An increase of relative speed rotates the time line t_R of any object away from the t_{0D}-heading from a common frozen starting point until it reaches completely a right angle at the relative speed of light c. The perpendicular set-up of four space-time axes is a mathematically and physically correct support construction that allows the relative time development of $t = 0$. It is important to realize that the grey areas of figure 1 are existent for every consecutive event in case of the generation of continuous time and inflation of space. The grey areas below the event horizons of individual length and time make the perpendicular construction mathematically and physically feasible.

- Trigonometric evaluations will be related to subjective and individual t_{0D}-lines, i.e. to the defined time of an original inertial frame of reference.

- If a reading of time will get a zero point at the start of each measurement, as usual in physics, this zero point means that the evaluation of time and length concerns already finished, i.e. completely rolled out Planck time intervals and an existing space length. Planck scale definitions are based on rolled out length and passed time developments. This observation of the past misses the quadrants of dark energies below the event horizons of length and time, which embed mass precipitations and regulate the roll out of length for the length-time quadrant above event horizon.

- $l_P = c \cdot t_P$ is valid with Planck's length l_P and with Planck's time interval t_P, $l = c \cdot t$ is valid, with prolongation of l_P to $l = k \cdot l_P$ and prolongation of $t = k \cdot t_P$. $k = 1,2,3,4, \ldots$ are integer. The letter c stands for speed of light in a vacuum.

- The total length of all four axes match each other, as it is possible to fully rotate all coordinates of a maximum inflated universe anticlockwise back to a situation before the big bang of the universe. The equal dimensions can be explained by the superposition of kinetic and potential energies and interacting opposing energy manifestations.

- The four quadrants that are separated by perpendicular space-time axes will be called "Relativistic Time Suite", abbreviated RTS, showing rotary super symmetry and an interaction on base of cause and effect.

- Space-time axes are summarized in some formulas by $t_{XD}(X = 0,1,2,3)$: They portray individual perceptions of time, length, statics and dynamics.

- A superimposed cross of four individual coordinate axes can be rotated in a frozen basic reference cross only with a sequential exchange scheme: $t_{0D} \leftrightarrow t_{1D} \leftrightarrow t_{2D} \leftrightarrow t_{3D} \leftrightarrow t_{0D}$.

- Development of the system time t_S of any observer shows $\Delta t_S \geq t_P$ with Planck time $t_P = 5.391 \cdot 10^{-44}$ s that limits the grey areas of figure 1 as an observer's individual event horizon of the progress of time, although the Planck time stays for any observer the same personal time interval.

- The construction with four axes is named "S-TQ" ("speed time quartet") and reflects the exchangeability of the four axes in a rotational symmetry caused by the individual perceptions of kinetic energies. The geometrical figure of a t_P^4-sphere across all four space-time axes that is characterized by the connection of all four quadrants in figure 1 in a point of space-time is named c_{T0}. t_P serves for all four axes because of the rotary aspects.

- In order to visualize the different space-time positions and developments in relation to $t_{0D}, t_{1D}, t_{2D}, t_{3D}$, it is possible to draw all four axes onto a plane, as shown in figure 4.

- This set-up causes opposing axes that are completely below the event horizons of each other, and the perpendicular axes at the event horizons.

- It is possible to differentiate the four axes by different impacts as space, or time, or static features with active speed of light, or dynamic features with passive speed of light. Figure 4 shows these classifications by the introduction of SP for a three-dimensional space-distance vector, TM for time, ST for a description of static features and DY for the description of accelerated processes that distort the space-time perceptions above the event horizon of an observer's individual length and time. The DY-line and the quadrant between DY and SP are revolutionary new aspects.

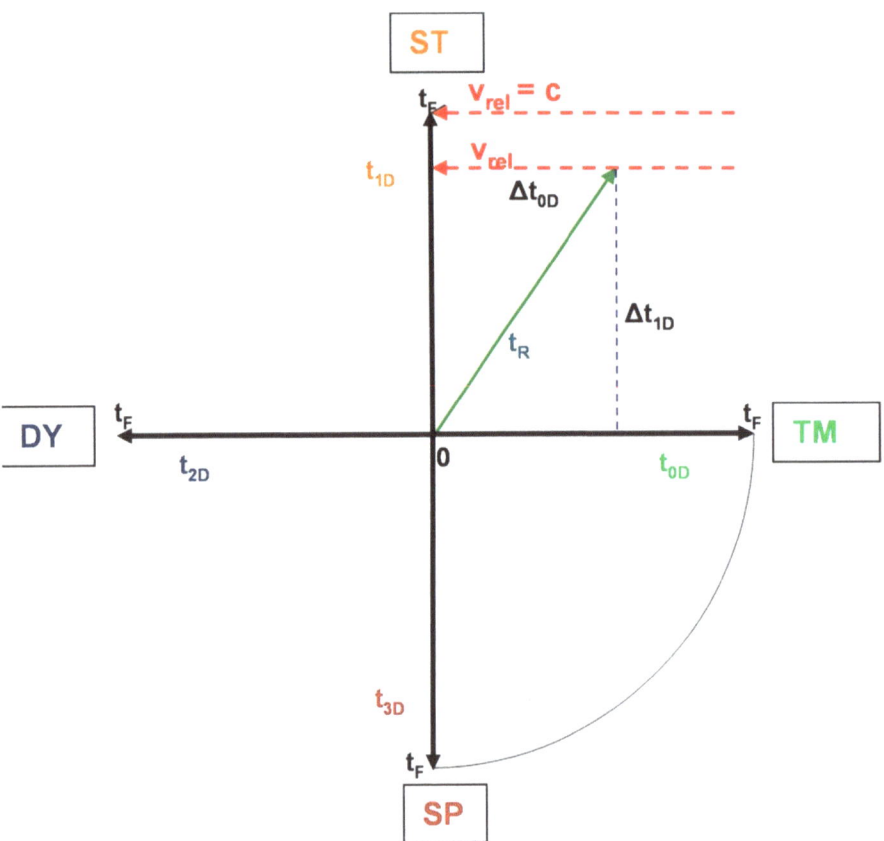

Figure 4: Relativistic dimension functions

The time function t_{0D} gets the name TM for time. The t_{1D}-function is called ST for its static storage features in time at active speed of light. The t_{2D}-function is called DY causing dynamic features at passive speed of light. The t_{3D}-function is named SP for three-dimensional space vectors from any location in space and time.

- Subjective evaluation on any t_R-line is always showing a right-angled, in relation to t_{0D} dilated S-TQ-set. This is a fundamental prerequisite for the constancy of speed of light and for a linear t_{1D}-speed scale from each t_R-line. This results in non-linear superposition of relative speed in case of adding speed vectors from the starting inertial frame of reference and speed vectors from relatively rotated frames of moving inertial systems. An individual perceived S-TQ-cross of an observer with own initiated relative speed or on any huge mass can be compared with up-winding of the spring of a mechanical watch with a cross formed key.

- Having an observation point above event horizon of length and time with sequential events on t_{0D} and passive speed of light on l_{2D}, we interpret the symmetrical picture differently: the four axes do not any more appear with equal parameters: t_{0D} and t_R appear as one-dimensional time in any definable point throughout space, t_{3D} shows up as a space length, t_{1D} is a speed barrier, t_{2D} appears with acceleration features.

- The gravitational constant G is experimentally derived with the value and unit:

$$G = 6.67428 \, (+/- \, 0.00067) \cdot 10^{-11} \text{N} \cdot \text{m}^2/\text{kg}^2, \text{ i.e. } \frac{\text{m}^3}{\text{kg} \cdot \text{s}^2} \quad \text{(Codata 2006)}$$

This gravitational constant G calibrates masses with unit kg within this arrangement, and, if applied to the relative acceleration value of t_{2D} only, with kg/m^2. The gravitational constant G links the values of space-time energy via the reduced Planck constant with the corresponding value of the standard model of physics expressed in Newton-meter Nm. It is valid:

$$l_P = \sqrt{\frac{\hbar G}{c^3}}, \quad l_P = c t_P, \quad l_P t_P = \frac{\hbar G}{c^4}, \quad l_P = \frac{\hbar G}{t_P c^4}, \quad l_P = \frac{E_P G}{c^4}$$

- A gravitational time dilation can be explained by reduction of the t_F-value within an S-TQ-diagram, as counter reaction to an increasing aggregation of masses. t_{0D} and t_{3D} face an equal impact in case of a relative t_F-contraction. t_{1D} builds up tension, but from a perspective within the t_{0D}-progress with constant relativistic ratio $\Delta t_{3D}/\Delta t_{0D} = \Delta t_{3D}'/\Delta t_{0D}' = 1$. This indicates that the speed characteristics in moving systems stay within this system just as they were without relative motion. Therefore, t_{1D} may be calibrated from any subjective t_{0D} timeline with linear development from 0 up to the speed of light for any chosen reference length of t_F and t_R. t_{2D}, however, shows its peculiar inversion from the viewpoint of time channel t_{0D}: The non-relativistic t_{0D}-view of Δt_{2D} as ratio $\Delta t_{3D} \cdot c/\Delta t_{0D}^2$ with unit m/s^2 and the scaled space-time-ratio $\Delta t_{3D}/\Delta t_{0D} = 1$ causes the inverted perception $\Delta t_{2D} \sim 1/\Delta t_{0D}$.

Result of all these considerations: the non-relativistic observations within the space-time reference frame with progressive t_{0D} can be transformed into symmetrical S-TQ-processes with four rotating space-time axes. The trigonometric evaluations have to consider reciprocals of opposing axes.

- References to theoretically or practically relative longer timelines t_{0D} are helpful to decide about dilation impacts of several connected moving scenarios on each other, according to interconnected history of objects and events, causes and effects. A longest timeline t_{0D} makes only sense in case of a one-way unbalanced picture of t_{0D}-caused t_R-effects.

- The differential geometry of any observable space-time scenario leads to different perceptions of space-time, if this scenario has been built up from t_{0D}, from t_{1D}, from t_{2D}, or from t_{3D}.

- Quantum leaps during segment transitions seem to cause deflections of energy components into new t_{XD}-directions with appearances and effects of superimposed vortexes, caused by residual momentum and detectable as electromagnetic waves and in form of spins and electrical charges of elementary particles. Electrical charges show static features of the ST-axis, quantized and decelerated from the speed of light of the ST-frame. Magnetic features of spins can be classified as quantized processes of the DY-axis origin and accelerated to quasi relative speed of light against the DY-inertial frame of reference.

- Stimulated, excited processes at event horizons of ST and DY disperse together in form of electromagnetic waves with active and passive speed of light features that are perpendicular to each other.

- Angular momenta of the anterooms of space-time quadrants appear as right-angled or rotary energy superposition within the observation space.

- The different angular momenta may initiate under certain circumstances an in-pair production of elementary particles, depending on the sequence of coupling and the reactions of the embedding space-time grid.

- The construction of baryonic matter with coupled angular momentum of the space-time grid leads to the matter waves that can be described by partial differential equations of second order in location and of first order in time: ST is derived by differentiation of TM, and DY by differentiation of ST, causing the second order in location.

- Figure 5 shows differently perceived +/- or neutral static momentum and -/+ or neutral dynamic angular momenta depending on the location of an observer and on the starting inertial frame of reference that initiated the production of the particle.

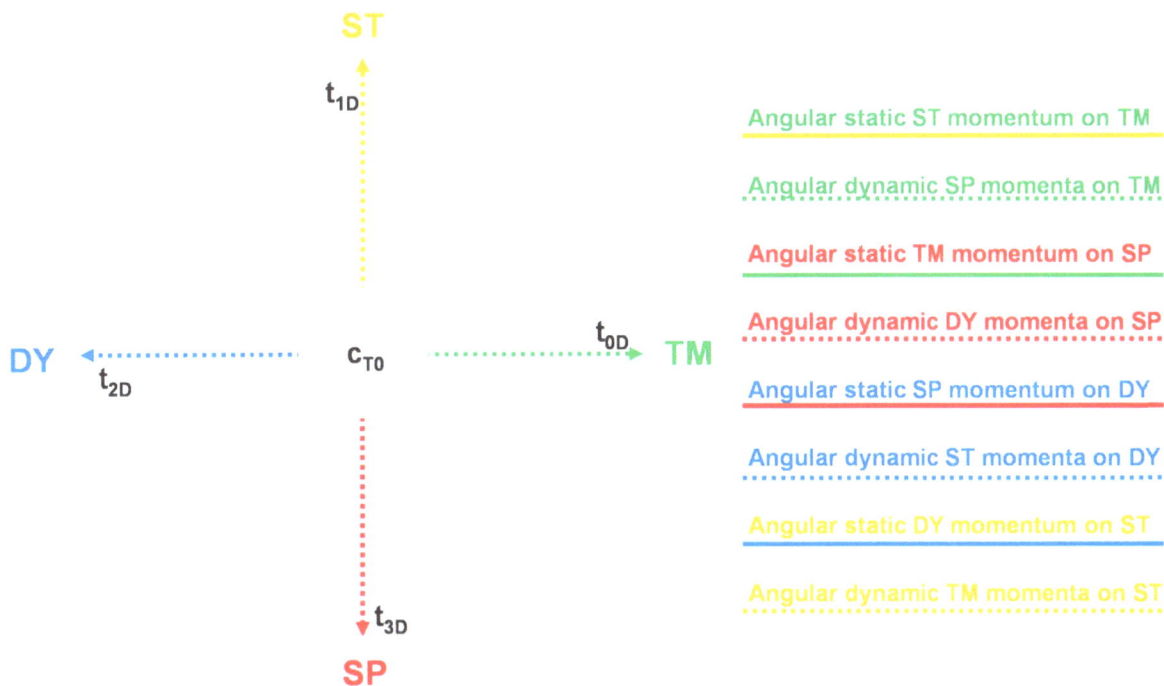

Figure 5: Static & dynamic angular momenta by active and passive speed of light

- Only relativity and the quantum mechanical break up at and below the event horizon of length and time allow four sequential axes that develop right-angled with rotary symmetry and that form a space-time grid below and at the event horizons of each other.

- The t_{0D}-development shows not a spatial dimension, but is developing in observation points throughout space as time parameter in any definable spot in individually experienced space, being the reason for its index 0D.

- Collective alignments of angular ST-and DY-momenta and of their spatial movements cause coupled, rectified electromagnetic fields. A generation of magnetic fields and electric fields are technical alignment applications.

- Relative speed means slowing down against t_{0D} relatively to observation posts by $-\Delta t_{0D}/t_{0D}$ and an increase of the t_{1D}-component by $+\Delta t_{1D}/t_{1D}$. This increase of t_{1D}-share requires a spatial movement describing kinetic energy or equivalent other energy impacts on the space-time grid as described by the general theory of relativity.

- The time processes on ST came to a standstill in relation to processes on TM, processes on TM in relation to processes on SP, processes on SP in relation to processes on DY, and on DY in relation to processes on ST.

- Electromagnetic ST-DY-waves can be transmitted at the speed of light without losses of superimposed information across long distances.

- Energetic, geometrical compensation points and areas are scaffolding for elementary particles and all space developments, forming a geometrical equilibrium quantum foam figure named "c_{T0}" for all definable space-time-spots above, at and below the event horizons for length and time.

- Trigonometric discussions of S-TQ-diagrams reflect the general theory of relativity for simple scenarios. Using absolutely equal scales for time, i.e. not considering any dilation, the trigonometric discussion of the ST-Q-diagram gets possible because the length of a time interval indicates the proportional relative energy content: A battery pack on any timeline will last exactly the same subjectively experienced time period, but the actual consumption of energy on rotated lines is slowed down in comparison with the consumption on the initial starting line because of time dilation.

  Example: The rotated Δt_R in the (+x, +y)-quadrant in figure 1 increases only the relative duration of energy of all sources in the moving object in relation to Δt_{R0D} and proportionally to the time dilation. Any motion or equivalent energy has an impact on an individual starting line t_{0D} and changes the space-time environment. These distortions can be captured for simple space-time scenarios by trigonometry in the rotary symmetry of space-time instead of Ricci-tensors and relativistic field equations. The reason for this simplification is the 360°-view of the space-time set-up.

- A rotating t_S-time and the perpendicular space-time cross of all four axes of the moving inertial frame of reference can be mathematically related to the starting constellation with axes t_{XD} by means of four cosine functions.

- The trigonometric formula for two ideally perpendicular time directions t_{XD} and $t_{(X-1)D}$ with t_{PXD} and $t_{P(X-1)D}$ of the Planck scale is:

$$t_{XD} + t_{PXD} = t_{FXD} \cdot \cos[z_{X-1} \cdot \arcsin(\frac{t_{(X-1)D}}{t_{F(X-1)D}})] + t_{P(X-1)D}$$

with $\quad \frac{t_{(X-1)D}}{t_{F(X-1)D}} = \sqrt{1 - \frac{v_x^2}{c_x^2}} \quad$ and $\quad \frac{t_{XD}}{t_{FXD}} = \frac{v_{XD}}{c_{XD}} \quad$ i.e. $\quad \frac{t_{(X-1)D}}{t_{F(X-1)D}} = \sqrt{1 - \frac{t_{XD}^2}{t_{FXD}^2}}$

$X = 1,2,3,4 \ (t_{4D} = t_{0D}), \quad z_{X-1} = 1,5,9,... = 1 + 4z_{S(X-1)}, \quad z_{S(X-1)} = 0,1,2,3...$

- The factor z_{X-1} in the trigonometric argument could have also the values of 2,3,4, 6,7,8,... This, however, is supposed to have a special impact on the space-time grid with the possibility to form elementary particles with different combined c_{XD}-vortexes and various types of local grid distortions by partial detached space-time areas. The *arcsin* has a chopper function.

- This formula is valid for all t_{XD}-quadrants because of the rotary concept and considers the t_P-zone below event horizon. t_P is changing direction together with t_{XDS} of the moving system S. The four $\vec{t}_P$ vectors on the four t_{XD}-axes are subject to consecutive perpendicular changes of directions, keeping space-time developments beyond the (+x, +y)-quadrant below the observer's individual l_P- and t_P-event horizons.

- These considerations indicate already the possibility to differentiate the four axes by relative flow and tension, as well as complementary aspects of slow down and acceleration in space-time.

- The *arcsin* argument has a range between 0 and 1 adjusting the *arcsin*-function between 0 and π/2: The counter z_{X-1} describes quantum leaps into adjacent locations in space-time beyond the given $t_{F(X-1)}$ frame.

- t_{XD}-time-dilations can be calculated with knowledge of the relative speed and gravitational time dilation can be analyzed by a decreasing $t_{F(X-1)D}$-reference interval with concurrent results but distinguishable intervention parameters on base of energy equivalence. $t_{F(X-1)D}$ and t_{FXD} are equal in value in case of equal space-time quadrant frames but acquired different functions because of their perpendicular orientation in space-time.

- Speed is increasing with $\vec{c}_{XD}$-direction on t_{FXD}, maintaining for the rotating t_{XDS} of a moving inertial frame of reference a perpendicular static speed of light direction: Static "Ante-room" is defined for TM and its adjacent TM-ST-quadrant [TM] as [ST], for ST and its ST-DY-quadrant [ST] as [DY], for DY and DY-SP-quadrant [DY] as [SP], and for SP and SP-TM-quadrant [SP] as [TM]. Dynamic "Follow-room" is for [TM] the impact area [SP], for [SP] the area [DY], for [DY] the area [ST] and for [ST] the area [TM]. Using vectors in trigonometric functions maintains the transparency and avoids complex mathematical rotary constructions and differential equations. This is only possible because of the rotational symmetry and the unchanged linearity of any rotating t_{XDS} cross of an inertial frame of reference based on the exchanges of relative space-time functions.

- A reference $\vec{c}_{XD}$ of static anterooms is stable for a linear speed increase because of the indifference that is based on the relative standstill of time.

- Initiation of speed of light processes seems to be only possible by harsh deceleration processes that cause the excitation of this static anteroom via the dynamic follow-room, releasing electromagnetic waves. The effect can be used to construct extraordinary strong lasers.

- Running accelerated through t_{XD}-equations, considering constant roll out of overall space-time but to perform additionally superimposed processes of oscillations by partially rolling up and rolling out space-time areas, it is possible to define phases of the four axes in relation to each other. The four space-time equations vary in their phases to each other: the [SP]-equation is out of phase by +90° or -270° in relation to the [TM]-equation, by ±180° in relation to the [DY]-equation and by +270° or -90° in relation to the [ST]-equation. These phases are important for the combination of the four basic equations and energy superposition.

- The passive speed level on any basic t_{XD} or t_{XSD} is the speed of light in relation to the perpendicular $t_{(X-1)D}$ axis or $t_{(XS-1)D}$. Individual evaluation on any t_{XD}-time will show an asymmetrical picture of interacting flow and tension energies, with slow down or acceleration, depending on the individual observer's event horizons of length and time: This space-time area of speed up processes at lower energy level explains many open questions in physics, for example the tunnel effect, where particles are suddenly able to overcome strict energy barriers by shooting this particles at these barriers.

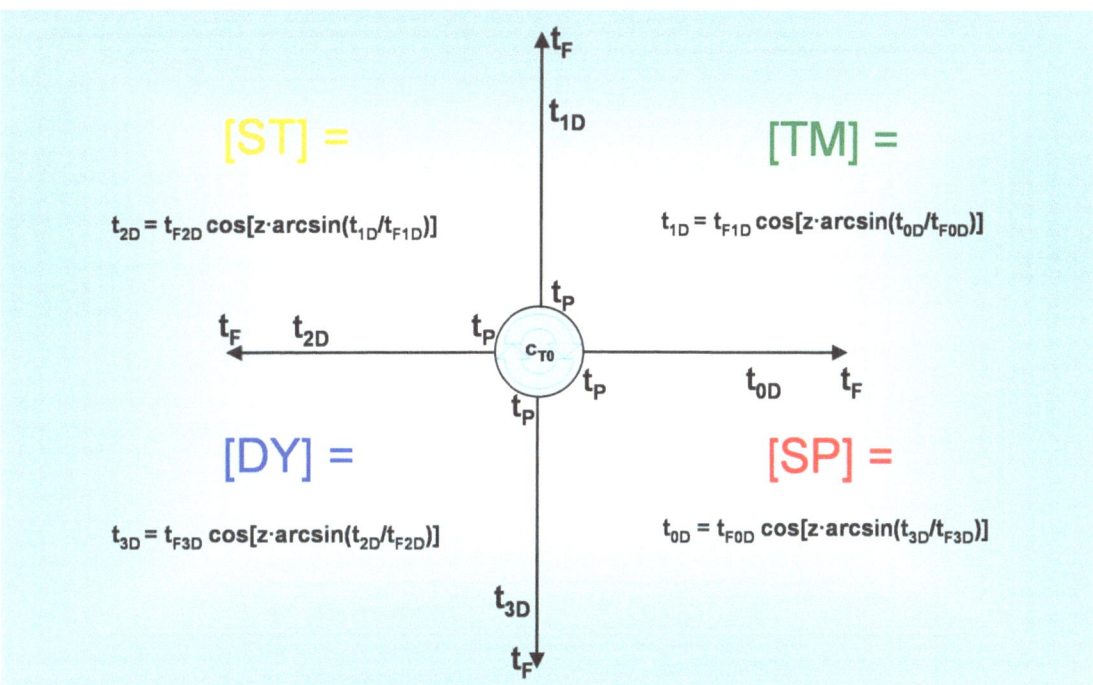

Figure 6: Basic t_{XD}-equations of all four sectors not yet related to each other

- Coupling of the four t_{XD}-equations causes processes of z-quantum phase leaping because of the system's component constraint $t > 0$ and the t_P-limits as event horizons of t_{XD} of each of the four axes.

- Linking of time functions with relativistic four perpendicular time directions leads to the following equation, with integer, quantum mechanical leap and twist-off-numbers z_{X-1} that are the basis for a detaching of particles:

$$E_Z = \left\{\vec{t}_{F0D} \cos\left[z_3 \cdot \arcsin(\frac{t_{3D}}{t_{F3D}})\right]\right\}$$
$$\times \left\{\vec{t}_{F3D} \cos\left[z_2 \cdot \arcsin(\frac{t_{2D}}{t_{F2D}})\right]\right\}$$
$$\times \left\{\vec{t}_{F2D} \cos\left[z_1 \cdot \arcsin(\frac{t_{1D}}{t_{F1D}})\right]\right\}$$
$$\times \left\{\vec{t}_{F1D} \cos\left[z_0 \cdot \arcsin(\frac{t_{0D}}{t_{F0D}})\right]\right\}$$

- Changing the sequence of cross product terms decides on the individual space-time construction phase. Note that all four cosine arguments are at $\pi/2$ if $t_{(X-1)D}/t_{F(X-1)D} = 1$ and 0 if $t_{(X-1)D}/t_{F(X-1)D} = 0$. The projection of linear changing $t_{(X-1)D}$-values on t_{XD} corresponds to the values of the set-up of the figures 1, 4 and 5 using relativistic transformation formulas.

- Changing t_3 into t_2 and t_1 into t_0 converts the cosine functions into sine functions and relative time dilations accelerate now the sine terms with t/t_P against t/t_F, turning tension characteristics into flow characteristics from the perspective of t_0 and time area TM: The resulting equation $E_{Zt0,2}$ shows the courses of time t_{2D} and t_{0D} squeezed into t_P-frames because of $\arcsin(t_{0D,2D}/t_P)$ and $0 \leq t_{0D,2D}/t_P \leq 1$. The split quantity $t_{(X-1)D}/t_P$ can be counted by $z_{1,3}$ factors. t_P is fixed for any t_F reference interval because of being the fixed space-time point of any relevance at the event horizon. z_{X-1} splits either into simultaneity spots in space (z_3) or delayed events in the course of time (z_1), depending on the appearance of the split axis.

$$E_{Zt0,2} = \left\{\vec{t}_{F0D} \sin\left[z_3 \cdot \arcsin\left(\frac{t_{2D}}{t_P}\right)\right]\right\}$$
$$\times \left\{\vec{t}_{F3D} \cos\left[z_2 \cdot \arcsin\left(\frac{t_{2D}}{t_{F2D}}\right)\right]\right\}$$
$$\times \left\{\vec{t}_{F2D} \sin\left[z_1 \cdot \arcsin\left(\frac{t_{0D}}{t_P}\right)\right]\right\}$$
$$\times \left\{\vec{t}_{F1D} \cos\left[z_0 \cdot \arcsin\left(\frac{t_{0D}}{t_{F0D}}\right)\right]\right\}$$

- The t_{2D}-axis is projected in a second step onto the t_{0D}-axis. Processes along t_{2D} distinguish themselves with inverted mathematical development in comparison to processes along the t_{0D}-axis. Therefore, inverted terms have to balance the E_Z-equation due to earlier findings of $\Delta t_{2D} \sim 1/\Delta t_{0D}$. Agreeing that the four trigonometric terms define their relativistic heading, all amplitudes t_{FX} may be cancelled to simplify the balance and impacts of asymmetrical observations from the SP-TM quadrant. This is possible in case of four equal quantities of t_{FX} as length, passed time, scaffolding maximum deceleration and balancing maximum acceleration.

The equation for a vector field $E_{Zt0R}(t \leq t_P)$ develops with four equal t_{FX} as:

$$E_{Zt0R} = 1/\left\{\sin_{0D}\left[z_3 \cdot \arcsin\left(\frac{t_{0D}}{t_P}\right)\right]\right\}$$
$$\times 1/\left\{\cos_{3D}\left[z_2 \cdot \arcsin\left(\frac{t_{0D}}{t_F}\right)\right]\right\}$$
$$\times \left\{\sin_{2D}\left[z_1 \cdot \arcsin\left(\frac{t_{0D}}{t_P}\right)\right]\right\}$$
$$\times \left\{\cos_{1D}\left[z_0 \cdot \arcsin\left(\frac{t_{0D}}{t_F}\right)\right]\right\}$$

- The t_P-acceleration terms of the DY-$\sin_{0D}$-contributions and the ST-$\sin_{2D}$ contributions change the perception considerably, especially having in mind that the DY-$\sin_{0D}$-quadrant shows up with accelerated processes and the ST-$\sin_{2D}$-quadrant with static features of relative frozen time, i.e. relative acceleration of the course of time in the TM-$\cos_{1D}$-quadrant. 0D, 1D, 2D and 3D describe space-time alignments of each trigonometric term in the respective $\cos_{XD}$-quadrant or $\sin_{XD}$-quadrant. Sin-quadrants and an opposing quadrant are below the event horizon of an individual observer's length and time.

- The E_{ZtoR}-equation shows a distinctive feature: Trigonometric arguments develop with $z = 1+4z_2$, $z_2 = 0,1,2,3...$ to maintain their positive directions $t_X, t_{(X-1)D} \geq 0$. These functions lose relevance in terms of length and time in case of an evaluation of the entire set-up out of only one space-time quadrant. Length contractions, time dilations and relativity of simultaneity of events can be studied as finished processes above event horizons of length and time. Below these space-time barriers any observer will notice uncertainty because of entire axis compressed on Planck scale. Crossing barriers leads to the complete exchange of the functions and space-time appearance. Quantum mechanics is based on these leaps and functional exchanges with Planck scale compressed uncertainty principle that can be partially captured by probability wave equations of quantum physics.

- The result of the leaping function at Planck scale is the construction of space-time quadrants in case of still differently oriented processes and tensions below the event horizons of length and time of an asymmetrical evaluation position. The overall view from outside shows then four space-time quadrants and the compulsion to interact in a rotary symmetry, if the rotation and the space-time roll out in clockwise direction leads to more stable, lower energy states, thereby increasing entropies of quadrants.

- Rotation of limited inertial frames of references in anticlockwise direction is possible on the expense of energetic countermeasures in space-time.

- The acceleration features of coupled trigonometric functions leads to the observation that space-time functions are sort of twisted off in t_P-tensor-field-slices by the following DY-function within the cross product equation.

- This consecutive t_P-twisting-off shows the effect that any t_P-leap causes a relativistic rotation of $4 \cdot \pi/2 = 2\pi$ into the original picture. This will be a starting point for the discussion of the function of the cosine argument multiplier z and its possible impact on detachments of space-time energy in form of elementary particles with related space-time distortion areas. This seems feasible because of the possibility to turn cosine functions by "space-time chopper factor" $z_{X-1} = 2$ by only 180° and then to combine the affected area by the four static and perpendicular c_{XD}^{-1} speed of light terms as detached c_{XD}^{-4}-constructions of space-time energy.

- Contributions of the $\cos_{3D}$-function are especially remarkable: t_P-leaps of the consecutive second twisting off process with the original starting base at the $\cos_{1D}$-function appear as LWH-perception with l_P-lengths, building up space with increasing expansion speed if observed across sufficiently long distances. The reason is the static behavior of the SP-quadrant in relation to the dynamic DY-quadrant and the damming up in between. The DY-quadrant and the (-x, -y)-quadrants are revolutionary findings.

- The coupled speed term c^4 with four different c_{XD}-directions fixed by the trigonometric term of this quadrant can be written into the equation:

$$\begin{aligned}
E_{Zt0R} = & \; 1/\left\{c_{0D} \sin_{0D}\left[z_3 \cdot \arcsin\left(\frac{t_{0D}}{t_P}\right)\right]\right\} \\
& \times 1/\left\{c_{3D} \cos_{3D}\left[z_2 \cdot \arcsin\left(\frac{t_{0D}}{t_F}\right)\right]\right\} \\
& \times \left\{c_{2D} \sin_{2D}\left[z_1 \cdot \arcsin\left(\frac{t_{0D}}{t_P}\right)\right]\right\} \\
& \times \left\{c_{1D} \cos_{1D}\left[z_0 \cdot \arcsin\left(\frac{t_{0D}}{t_F}\right)\right]\right\}
\end{aligned}$$

- Only the quotient v/c_{0D} is relevant for any relative speed in the [SP] space quadrant between axes SP and TM, i.e. t_{3D} and t_{0D}.

- Starting processes from space-time positions that are not coinciding with t_{0D} leads to the relative dilated system time t_S of such positions. It is the reference time t_S of an inertial system or event or process in this system:

$$\begin{aligned}
E_{Zt0R} = & \; 1/\left\{c_{0D} \sin_{0D}\left[z_3 \cdot \arcsin\left(\frac{t_S}{t_P}\right)\right]\right\} \\
& \times 1/\left\{c_{3D} \cos_{3D}\left[z_2 \cdot \arcsin\left(\frac{t_S}{t_F}\right)\right]\right\} \\
& \times \left\{c_{2D} \sin_{2D}\left[z_1 \cdot \arcsin\left(\frac{t_S}{t_P}\right)\right]\right\} \\
& \times \left\{c_{1D} \cos_{1D}\left[z_0 \cdot \arcsin\left(\frac{t_S}{t_F}\right)\right]\right\}
\end{aligned}$$

- An E_{ZtoR}-equation shows that t_0/t_P-factors of relative time developments and relative spatial movements accelerate processes up to passive and active speed of light because of the rotary set-up. This leads to the new interpretation of all electromagnetic processes, especially the magnetic features and the spins of free and nuclear particles.

- Advanced mathematic calculation methods are able to describe multiple interacting E_Z-constellations. Superimposed fields can be visualized and calculated with Einstein's differential operations but enriched with three space-time quadrants and their rotational event horizons at Planck scale.

- Observing E_Z- balances with quantum mechanical leaps and considering the rectangular set-up simplifies the E_Z-equation, using an E_Z-quotient:

$$E_{ZtoR} = \frac{\{c_{1D}\cos_{1D}[z_0 \cdot \arcsin(\frac{t_S}{t_F})]\} \cdot \{c_{2D}\sin_{2D}[z_1 \cdot \arcsin(\frac{t_S}{t_P})]\}}{\{c_{0D}\sin_{0D}[z_3 \cdot \arcsin(\frac{t_S}{t_P})]\} \cdot \{c_{3D}\cos_{3D}[z_2 \cdot \arcsin(\frac{t_S}{t_F})]\}}$$

- E_{ZtoR} gives an overview about streams, static tensions, accelerations and decelerations, including considerations of mathematical term inversion. Using the summarizing terms [TM], [ST], [SP], [DY] leads to transparent relations of individual E_{ZtoR}-constellations from a frozen observation spot in the SP-TM-quadrant, provable by electromagnetic processes:

$$E_Z = \frac{[TM] \cdot [ST]}{[SP] \cdot [DY]}$$

- Exchange of static, opposing and dynamic space-time energy between [TM], [ST], [DY] and [SP] at and below event horizon ensures continuous t_P-patchwork. The appearances of positive and negative polarities are a result of in-pair production of static charges and of an acquired spatially effective spin in a three-dimensional space-environment.

- These polarities can be experienced by two opposing [SP]-[ST]-charging features, with point charge characteristics in [SP]-space, and as coupled [DY]-[TM]-processes of magnetism with passive speed of light features in SP with respect to magnetic flow or detached magnetic momentum of [DY]. As important result, attraction and repulsion scenarios are possible in three-dimensional space for electricity and magnetism, and screening off of these forces. A shield is not possible for the opposing (-x, -y)-area of gravity in the (+x, +y)-quadrant.

- The separation into positive charges and negative charges determines a spatial rotation direction of coupled magnetic fields in case of relative speed, according to the rules of electrical engineering.

- This movement is forced because of t > 0, and it is slowed down because of four static sliding bases. A relative velocity causes an overlaid rotation into the opposite direction. A starting point of constant relative speed is defined as zero reference for individual time and length comparisons that will always show system immanent increase of space, despite of storage in multifarious forms of acceleration densities and speed scenarios.

- The magnetic field constant μ_0 and the electric field constant ε_0 with the permeability μ_r or permittivity ε_r, respectively, describe ST – DY – SP – TM interactions and observations without and with masses. Fixing the magnetic field constant:

$$\mu_0 = 4\pi \cdot 10^{-7} \text{ (V·s / m·A)}$$

by definition and rotary magnetic tension (Volt V indicates electric tension and Ampere A the generation of magnetic tension) results in an electric field constant:

$$\varepsilon_0 = \frac{1}{\mu_0 c^2} = 8.854187817 \cdot 10^{-12} \text{ (C / V·m), for scaled } |c| = 1: \ |\varepsilon_0| = \frac{1}{|\mu_0|}$$

The formula of μ_0 confirms the relative inverse contributions of SP (→m) and DY (→A), and the linking of ST (→V) and TM (→s) in the numerator, and SP and DY in the denominator. The application of $1/c^2$ takes the rotation across two space-time axes into account.

- There are four possible perspectives for the energy balance. The rotated constellations of these four individual views are energetically equivalent:

$$E_{ZR} = \frac{[TM] \cdot [ST]}{[SP] \cdot [DY]} \quad E_{Z1/TR} = \frac{[SP] \cdot [TM]}{[DY] \cdot [ST]} \quad E_{Z1/R} = \frac{[DY] \cdot [SP]}{[ST] \cdot [TM]} \quad E_{ZTR} = \frac{[ST] \cdot [DY]}{[TM] \cdot [SP]}$$

V) Gravitational time dilation within the super symmetry

The superposition of space-time positions of events leads to very complex processes, because of the interaction of the involved space-time quadrants and forces, with limited freedom of the movements through space and time, and bounds of collective system compulsion. Increasing mass aggregations oppose basic time, with the impact of increasing slowdown of relative time development, however, staying in the present of relatively accelerated time processes. If a light wave leaves a large mass, an observation shows the effect of a red shift. This effect describes a gravitational dilation effect.

The time t_{0D} progresses on a larger mass slower than on the remote and smaller mass, although they stay in the presence of each other's time. The time ratio $\Delta t_{0D}/t_{FM}$ which is evaluated for the larger mass is higher than the time ratio $\Delta t_{0D}/t_{F0}$ of the remote object, taking an equal Δt_{0D}-pace of a t_{0D}-clock for both locations. Recalculating these ratios with equal references t_F the Δt_{0D}-length of the oscillation period in departing objects get relatively longer and the object's time runs relatively faster, though staying visibly in the present time of the starting point on the large mass.

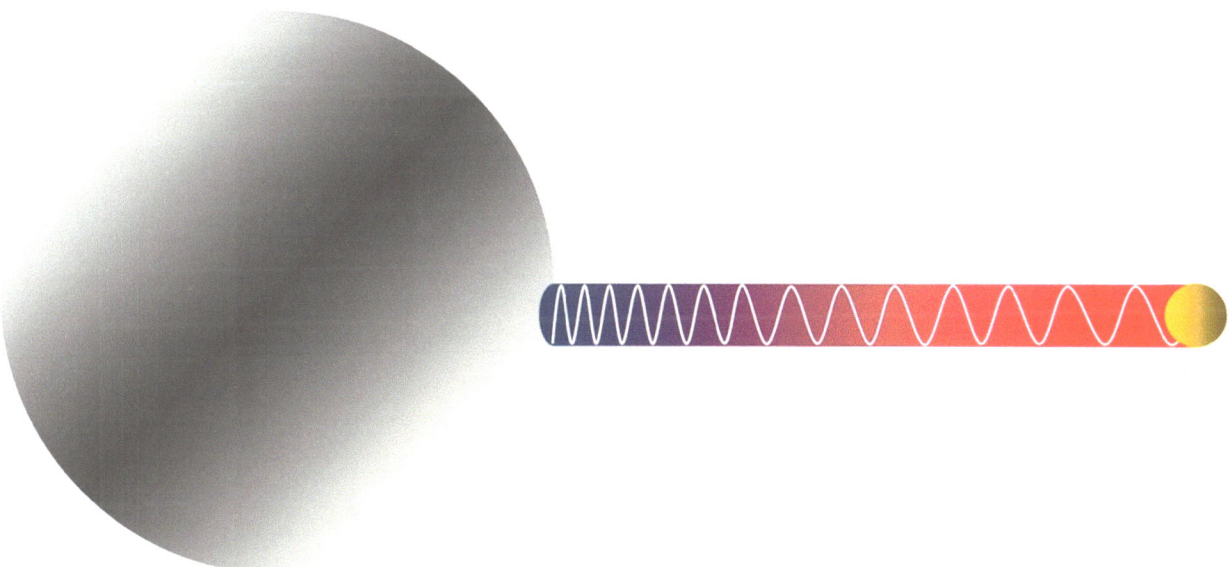

Figure 7: Gravitational red shift

If light is leaving from a large mass aggregation, it loses relative t_0/t_F-energy. Therefore, the light wave gets relatively longer. All objects leaving the gravitation field of a large mass decrease their t_{0D}/t_{F0}-ratio against the t_{0D}/t_{FM}- ratio of the large mass, due to increasing t_{FX}-resultants. This indicates the reduction of relative space-time energy density and the relativistic energy storage capacity of the space-time grid.

Gravitational dilation describes a picture of space-time embedded masses with respect to one common t_{0D}-time axis, with $v_{rel} = 0$, and the t_{0D}-starting time of measurement with $t = 0$. Visualization in the TM-ST-DY-SP-diagram is similar to figure 1: The measurement of gravitational time dilation with an observer's seeming continuously equal pace of clocks is now related to t_{0DM}/t_F as the defined zero point of any relative measurement, with $v_{rel} = 0$ throughout the diagram. The object's time t_{0D0} does not rotate like in figure 8. The time line t_{0DO} in figure 8 serves for comparisons with figure 1.

Figure 8 shows the possibility to express time dilation that is caused by gravity with $t_{F1,2,3...}$-contractions or corresponding rotations of coordinates, but now of the time line of the large mass, due to the equivalence principle of mass energy and corresponding kinetic energy.

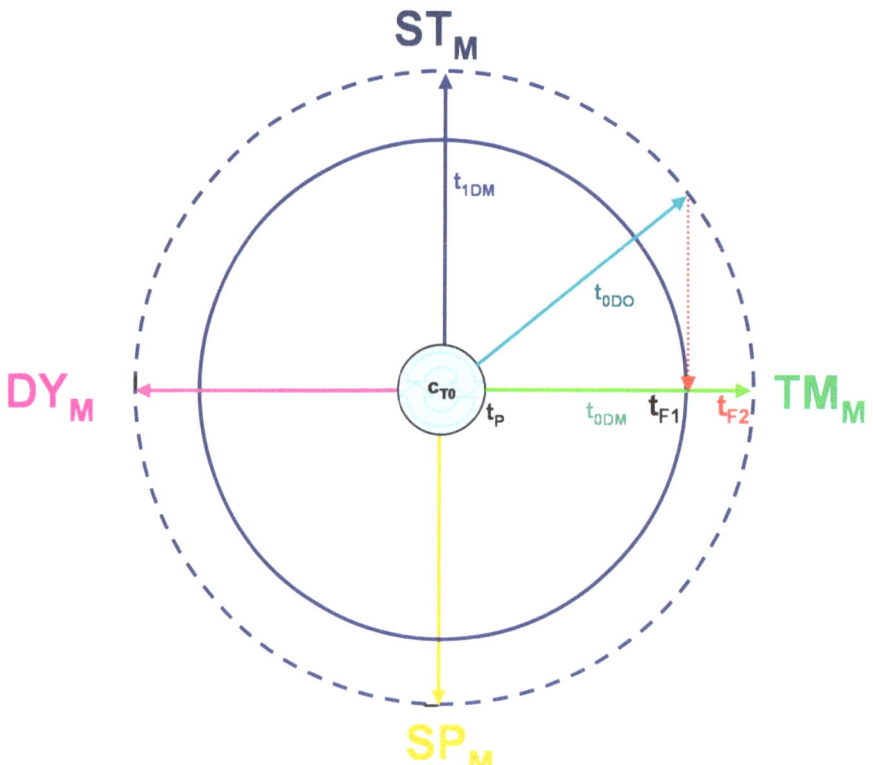

Figure 8: Gravitational dilation

Mass aggregations cause opposing space-time reactions, with individual values of t_F. Change of t_F causes relative contractions of space-time frames indicated by different circumferences. Relative velocity, however, rotates the t_{0DO}-vector of an object. The projection on t_{0DM} of the large mass (red arrow) shows relative dilation of time in the moving object with respect to measurement reference t_{F1}: time dilation by relative speed and gravitational time dilation are absolutely equal at this point with respect to reference t_{F2}.

Comparisons of individual masses are based on t_{0DM}/t_F-ratios and on the corresponding space-time-frame impacts. Figure 8 combines time dilations by relative speed and gravitation with equal impact, visually differentiated. It shows how space-time density is increasing with these two parameters, increasing or decreasing relative energy storage. The decisive difference between gravity and the other three fundamental forces (weak radioactive nuclear interaction, strong nuclear interaction and electromagnetism) is the necessity of a double rotation across two space-time axes to interact with or via the opposing space-time quadrant. This interaction is restricted and can only be described at the Planck scale with opposing space-time quadrants. Therefore, gravity is by far weaker than the other three fundamental forces of nature, cannot be shielded and shows no polarities in LWH-space.

VI) New derivation of gravitational constant G and masses

The following reflections upon the gravitational constant G and masses of leptons and bosons are based on the superposition of distinguishable E_Z-scenarios and 8 different kinds of speeds of light parameters, if we allow the detachment of quantized space-time areas by influences of trigonometry, adding 4 superimposed 180°(- c^{-1}) terms to 4 basically (+c) t_{XD} directions.

Asymmetrical appearance of the overall grid from the (+x, +y)-quadrant with length l [m], time t [s], speed v/c [m/s] and acceleration a [m/s²] for the four axes leads to the following vector product of space-time units of the sequentially perpendicular axes construction: $\vec{m} \times \vec{s} \times \overrightarrow{(\frac{m}{s})} \times \overrightarrow{(\frac{m}{s^2})}$. This is our distorted perception of the symmetrical construction that leads to a mass by cutting out a limited area and turning it 180°. Because of the sequentially perpendicular unit vectors, the result of the vector product is $\frac{m \cdot s \cdot m \cdot m}{s \cdot s^2} = \frac{m^3}{s^2}$. The difference of this vector product to the basic relationship of tension has to be proportional to the produced mass M, with gauge unit kilogram kg, i.e. $\Delta \frac{m^3}{s^2} \sim M\,kg$, or, $\frac{\Delta \frac{m^3}{s^2}}{Mkg} = G = constant$. Therefore, the unit of the gravitational constant G develops to $\frac{m^3}{s^2 kg}$. The value of this constant G depends on the chosen kilogram reference mass and can be calibrated in a gravity field by gravity force $F = G\frac{m_1 m_2}{r^2}$ in $N = \frac{kgm}{s^2}$ between two spherical masses m_1 and m_2 at a distance r between their centers. Rearrangement of the G-equation and adjusting it to the possible l_P-maximum $M_{max} [kg] = \frac{l_P^3}{t_P^2 G} = \frac{c^2 l_P}{G}$ leads to $M_{max} = m_P = 2.17644 \cdot 10^{-8} [kg]$, with Planck length $l_P = 1.616252 \cdot 10^{-35}$ m. Mass m_P is the Planck mass m_P, describing space-time limits of a mass concentration: $m_P = \frac{E_P}{c^2} = \sqrt{\hbar \frac{c}{G}} = 2.17644 \cdot 10^{-8}$ kg showing still the crossing of two space-time axes that generated the c^2-term of $\frac{E_P}{c^2}$, i.e. the rotation by 180°: The energy equivalence of baryonic masses $E=mc^2$ gets its rotary space-time characteristics by c^2. Speed of light scaling of time leads to c=1 and requires special attention on the different c-directions.

$E = mc^2$ cannot reflect the entire process, because of a rotational symmetry. Energy E is the measure of the space-time cross against the rotated mc^2, requiring another c^2-term keeping the symmetry and energy balance across four perpendicular space-time axes with symmetry m^4/s^4 and a maximum of c^4 of four different speed of light directions. Any mass compensates this balance with $\Delta\ s^4/m^4$ and maximum c^{-4}, being a 180° rotated limited area.

E_z-equations describe t_p-phases of events: Quantum mechanical number z determines energy detachment, if electrons are supposed to be released by a 180° turn of a quantized space-time area, doubling the arcsin function by z=2, pulling the cos-function fully into the negative sector, with theoretical $\frac{t}{t_F} = \frac{2 t_P}{t_P} = 2$ of the arcsin arguments, staying practically at the mathematical maximum of $\frac{t}{t_F} = \frac{t_P}{t_P} = 1$. This seems feasible because of opposing and then deflected and cut out speed of light components. There are 4 static c^{-1}_{XD}-factors in any E_z-equation with the 4 different (+c) orientations of t_{XD}-vectors that may be combined in multifarious ways if such distinguishable stable E_z-scenarios can be separated and superimposed. Scaling of time with speed of light leads to 8 different unit vectors $\vec{1}$. The time interval of at least $2t_P$ is needed for the arcsin cut out and the 180° cos-rotation of quantized space-time areas, ending up in the generation of elementary particles. This $2t_P$-cut out and rotated area is split into the free pair of electrical +/- particles, an electron and a positron, or, 90° rotated, electromagnetically neutral pair of particles, an electron-neutrino and anti-electron-neutrino, speculatively. If tidal energies are compressed into a nucleus, the resulting mass should be amplified by c^4-leverages. The relativistic mass increase is generated by an additional space-time grid distortion that shows these c^4-leverage impacts. Calibrating the smallest feasible electrically charged free mass particle with

$c = 299792458$ m/s, force $F = m_F \cdot a$, imposed acceleration a and equation $m_F = \frac{F}{a} = \frac{X}{c^4} \cdot C_A$ ($C_A = 7.358202 \cdot 10^3$ kg $\cdot$ m^4/s^4) shows amplification factor X=1 related to the rest mass of an electron: $m_F = m_e = 9.109382 \cdot 10^{-31}$ kg. The rest mass of a proton could be described by $m_{Pr} = \frac{X}{c^4} \cdot C_A$ for two up-quarks and one down-quark and stable c^4-leverage multiplier X =1836.153 derivable from combined distortions of two +2/3 e charged up-quarks and one -1/3 e charged down-quark, forcing only in combination the respective m_{Pr}-c^4-leverage position of the proton between the four space-time axes. Term $R_A{}^{-1}{}_{min} = G \cdot a_{mx}$ [$\frac{m^3}{kgs^2} \frac{m}{s^2} = \frac{m^4}{kgs^4}$] with superimposed grid acceleration a_{mx} on a mass leads to maximum c^4/m_X [$\frac{m^4}{s^4 kg}$] and $R_{A\,min} = m_X/c^4$ [$\frac{kgs^4}{m^4}$]. This acceleration of any mass reveals the space-time symmetry by inertia. The electron's rest mass gauge in the m_F-equation above sets m_X= 1 and shifts the unit kg into C_A to better reflect a pure amplifying $\frac{X}{c^4}$-distortion of the grid. The various quantized energy portions are described by particle physics.

VII) Expedition through the standard model of physics

Descriptions of nature with physics are orientated towards and tuned with our human perception faculties. All measurements register defined physical quantities that capture qualities and characteristics of objects. An observed object may be a concrete object, but could be also any physical state or a process. The conformity to all physical laws is expressed by mathematical descriptions and logical linking of all physical quantities and processes. All quantities are always described by a value and a connected unit.

Physics defines seven basic types of quantities. These are length, time, mass, temperature, electric current strength, substance quantity, luminous intensity. Other types of quantities can be derived from these basic seven types either by formulations of natural laws or, for specified purposes, with products and quotients. An RTS is based on a fundamental quantity only, tidal energy, with the system's acceptance of a relative change of flow and tensions that is perceived either as space lengths, as acceleration forces, or in various forms of condensing matter. All other types of quantities and all natural laws can be derived from a rip into tidal flows and tension energies. An S-TQ-model of rotating interacting relativistic relative time developments opens revolutionary and complementary perspectives on all kinds of energy manifestations in nature. The descriptions of these processes extend the ways of looking into the details of physics, which is quite helpful to bring all natural phenomena into an overall context and down to and even far below Planck scales and in spite of asymmetrical appearances of space-time.

In the course of this chapter, we just skim through the disciplines of physics, making cross references to compatibility with the RTS. These six disciplines are mechanics, thermodynamics, acoustics, optics, electrics, and nuclear physics.

Mechanics

Let us start with mechanics. The basic length to measure distances is today one meter [m]. It is the length, which light in a vacuum covers in the time interval of 1/299 792 458 seconds. Any length of the relativistic time suite can be considered as a manifestation of a subjective multi-reactive power patchwork and a derivative quantity of Planck energy and the gravitational constant that links space-time energy with mass-based energy. The interval of one second [s] is basically a defined quantity, based on the radiation of cesium 133. However, the Planck time is a derivative quantity coming out of the Planck energy, the gravitational scaling constant G and speed of light c.

The mass unit "kilogram [kg]" is a defined quantity, and related to a regular mass of a kilogram reference prototype in Paris, made of platinum-iridium. The cylindrical reference mass has a height of 39 mm and a diameter of 39 mm. The rest mass is supposed to be constructed by c^{-4}-elements of the space-time quadrants. These are neutral to the value of a rest mass if c is scaled to c=1. The relativistic mass is relatively growing with increasing relative speed, proportional to the time dilation. In case of superimposed three-dimensional t_{3D}-vibrations that are caused by extreme temperatures, the dimensions and constructions of matter become blurred (plasma).

The difficult part of re-thinking needs to be invested in the basic discipline of physics, namely kinematics. Two factors have to be considered carefully. On one side the movement on defined and only theoretically straight lines, if acceleration and reacting forces have to be described with application of integral calculus and with the quadratic speed terms. On the other side, it concerns the use of differential calculus, formally eliminating dimensions in case of using straight lines. Space-time-speed-acceleration diagrams reveal that every process should be calculated with a rotary space-time process. Integral calculi of kinematics recover reductions of dimensions, in case of using straight motion lines in a system with length, width, and height, LWH. Therefore, all processes at a sufficiently low relative speed can be usually captured with acceptable high precision.

Because of the relativistic three-dimensional spreading of t_{3D}-directions and t_{0D}-spots, all length developments in LWH are straight throughout the visible mass-free universe into all radial outbound directions from any observation point throughout the universe, with an increasing retrospective view. The rotary construction of an energized space-time explains the impression of a central location of each observer throughout the universe.

Comments to Newton's three, non-relativistic axioms:

Axiom 1:
"Without external forces, any mass remains in the state of rest, or in the state of straight and steady motion". This characteristic of masses is called inertia. There are always gravitational forces involved, but with low impacts across enough long distances. The t_{1D}-tension factor is not considered in this non-relativistic picture of inertia, and thus no relativistic mass increase. The reason for the validity of this axiom for non-relativistic descriptions is the perpendicular quasi reactive energy construction of length and time that keeps the length and time axes for an observer in his own system always at a right angle. Chapter VI revealed the space-time mathematics of inertia.

Axiom 2:
"Effective forces from initiated accelerations are always proportional to each other: $F \sim a$". Looking at the dynamic super symmetry, this fact becomes obvious. Each aligned force on a mass causes the shift of balanced forces between t_{0D} and t_{2D}, which has to be transcribed with a superimposed t_{2D}-acceleration, holding against the relative t_{0D}-developments. This process is quantized by the rotary energy set-up at event horizons, as well as gravity. The decisive difference between gravity and the other three fundamental forces (weak radioactive nuclear interaction, strong nuclear interaction and electromagnetism) has been identified as limited interaction possibility with the opposing space-time quadrant via the adjacent quadrants. Therefore, gravity is much weaker than the other three fundamental forces of nature, cannot be screened off and shows no LWH-polarities. The $1/c$-elements of any mass are so to speak in the wrong, opposing space-time quadrant,

trying now to make their way home via the short cut of mass defects, or, on the long run, via gravity and the critical mass aggregations of black holes.

Axiom 3:
"If a mass body has a force impact on another body, it experiences from the other body an opposing force with equal strength". The reason for this is the function of the involved c_{T0}-compensation-points, causing all action-reaction processes of individual S-TQ-space-time-mass interaction scenarios above quantum thresholds.

Nevertheless, these three axioms of Newton capture only a small part of the processes within the entire system: they cannot explain influences of tidal vortexes and elementary angular momentum, and the effects of collective alignments. These axioms describe t_{0D}-t_{1D}-t_{2D}-t_{3D}-interactions at low speed.

Relativistic mechanics allows time corrections in high speed scenarios with the formulas of the SRT in low gravity environments. However, any clock on earth runs already approximately by a factor 6.95317×10^{-10} slower than in far space that is characterized by nearly no presence of masses.

The proportion between the mass of a defined body and its volume is called density. Cohesion and adhesion forces, causing the density of a body, react with increased t_{3D}-vibration in case of increasing the temperature, or in case of compression. Gaseous materials show up with temperature and pressure depending densities. In order to enclose a gas, there is the necessity of an opposing pressure that could be realized by appropriate materials with a sufficiently higher t_{3D}-compression or by interplay of cooling and gravitation, just like in the earth's atmosphere.

The springiness of mechanical springs demonstrates an elasticity of matter manifestations. The construction of elastic matter requires special material structures, allowing the stretching of material t_{3D}-distance with simultaneous t_{3D}-compression of other areas. This process initiates an increasing t_{3D}-vibration, causing the partial increase of material temperatures.

Within gravity fields of large masses it is only possible to move with at least one component that is opposing this force field. Aerodynamic locomotion, rocket propulsion and buoyancy of hot-air balloons are different examples of such opposing forces. But they all have one feature in common: their individual t_{3D}-structures of substance densities support lift, buoyancy, or propulsion in favor of gaining or maintaining a relative LWH-t_{3D}-distance. All processes of motion are connected with an increased t_{3D}-vibration of parts and materials that are involved in the repulsion techniques to gain a relative distance. The rotary symmetry of space-time builds the base for propulsion and buoyancy.

Any form of energy in the universe is in principle kinetic energy, even if energy processes are relatively extremely slowed down. Because of the S-TQ construction and via residual angular momentum, there is the possibility to store kinetic energies in different forms of long-term potential energies, which can be released in case of a necessity. Release of compact matter manifested energy is possible, but with an involvement of extremely high distortion forces, usually triggered by implosions, collisions, or compression.

Chemical states of substances are mainly a function of the temperature, in combination with various affinities to link and interact with other substances. Between molecules there are interdependent acting forces, responsible for chemical states and in case of solids and liquids responsible for volumes. The rotary symmetry adjusts an equilibrium distance between molecules.

The range of cohesion forces between the molecules of a body is low and the repulsive force decreases faster than the attracting force. The resulting sphere of interaction has the radius of approximately 10 nanometers [nm]. The cohesion of molecules of a substance and an adhesion of molecules of different substances can be designed and controlled by nanotechnology, in order to interact in more efficient ways. All products and processes can be tailored to applications and many innovative inventions are about to come that use the transparency of the rotary set-up of an extended space-time.

Mechanical oscillation is a periodical spatial shift of masses. All initiated t_{1D}-processes are captured by speed, t_{3D}-processes by length measurement, being two quite different manifestations of tidal energy and its storage.

Thermodynamics

Thermodynamics differentiates between thermal state and thermal energy. Thermal state takes temperature as summarized description of microcosmic t_{1D}-t_{3D}-vibrations of all elementary particles inside the atoms, of combined atoms or molecules, either free in space, or fixed by grid constructions.

For thermal energy, a principle of energy conservation is valid, too. Thermal energy can be compared with law conformities of mechanics, because it is merely motion energy of elementary particles, of atoms, and of molecules. Aggregation of matter reduces possibilities of three-dimensional movement inside and outside of gravity fields.

In return for these restrictions, it is possible to initiate motions in any radial direction of three-dimensional space, if there is any physical possibility or a technical solution to push off from other masses, initiating propulsions in opposing directions. If a mass becomes extremely large in relation to other masses, the large mass absorbs forces of pushing, like our earth does.

Acoustics

Sound generation and dispersion are superimposed forces, generated by low frequency oscillations and waves, and distributed by suitable carriers. Applying only low amplitudes of sound oscillations shows measurement results of a dispersion speed that mainly depends on mechanical properties of the carrier, not on the frequency of the sound wave.

Any relative speed between a transmitter and the receiver causes the wave distortions, known as the Doppler-effect. Increasing constantly the distance from the receiver with respect to the observation at the receiver, initiates the relative stretching of the sound frequency of the transmitter, resulting in a lower frequency observation at the receiver. If on the contrary a transmitter approaches the receiver, the waves appear relatively compressed with the effect of the relative increase of frequency. The Doppler-effect differs from relativistic processes, because the basic speed of tidal energy is always speed of light that can relatively fall only below this original value in case of a rotary process in space-time. Therefore, STR-dilation is observed with any heading in space, i.e. inbound, outbound, or passing by. The Doppler-effect is the result of a translation of interacting t_{1D}- t_{3D}- points in space.

Optics

Light shows combined ST and DY-features, being an electromagnetic wave and a particle stream without rest mass, depending on the strength of an excitation. Light follows t_{3D} but from its perspective rotated into t_{0D} with its time t_{0D} rotated into the full t_{1D}-stagnation of time. Waves and particles show contrary space-time characteristics, building the base for peculiarities of the observations of light interferences: The evaluation of light interferences differ depending on the observation as a particle stream or as a wave. This physics riddle is solved by combined contrary space-time features of light.

The interactions in atomic grid structures of a medium reduce the speed of light, seemingly. This effect is triggered by microcosmic processes, if light is sent through any medium. Nevertheless, the effect can be still studied in a rotational extended space-time diagram: Reduction of relative speed can be related to a relative rotation from t_{1D} towards t_{0D} and from t_{3D} towards t_{2D}. t_{2D} causes refraction of a slanting ray in the denser medium towards the plumb.

Electrics

Electricity is a description of space-time features that appear as interactions of quadrant ST with quadrant DY by means of electrically charged particles. Motions of electrons generate electric current and rotating magnetic fields. This kind of ST-DY-excitation is bound by existence of SP- and TM- matter elements to trigger and transfer the interaction. Electric current experiences electrical resistance in materials, causing molecular t_{3D}-vibrations that result in increasing material temperatures in comparison with the environment.

A possibility to reduce the electrical resistance of electrical wires is cooling down close to the absolute zero level T_0. The result is the superconductivity effect with very low resistance because t_{3D}-vibrations are figuratively frozen if the temperature gets close to absolute zero T_0 and electrons experience low resistance across middle way length. Specialized high-tech laboratories work on nanotechnological structures that provide superconductivity effects already under environmental conditions of earth's surface.

Because of the transverse tidal constructions, electric field lines are always leaving perpendicularly to the surfaces of conductors, and, according to the definition, disperse from positive poles towards negative poles. Movements of electrons within an aligned electric field are perpendicular to a generated rotating magnetic field, showing the first axes crossing by the right angle and a second crossing by the rotation of the magnetic field. Electromagnetic interactions can be synchronized with receivers throughout the spreading range of electromagnetic waves. Without the angular tidal momentum of masses any synchronization, like in antennas, would be impossible.

Atomic and Nuclear Physics

Atomic and nuclear physics describe various forms of elementary particles and formations and interactions within atomic structures. The particles have been summarized in the standard model of particle physics. A missing link is the experimental confirmation for the actual appearance of a mass which is currently assumed to be provided by energy exchange of a Higgs particle with a mass background field. The rotary model of space-time presents the (-x, -y)-field as a mass background field for the perception of masses and the rotation and detachment of quantized energy as a feasible concept for the appearance of the various forms of particles of the standard model.

Dimensionally reduced models cause the necessity for constants of nature, postulated energy relations, and super symmetrical particles like neutralinos that are todays candidates for dark matter. In the rotary space-time model super symmetrical particles could be defined below event horizons of length and time in the (-x, -y)-quadrant as theoretical patterns and clusters of tidal reversed energy.

The rotary model of space-time simplifies comprehension and assessments of energetic set-ups and processes. It portrays processes in space-time in a way that any neutral observer would perceive it in a monitoring position that is completely outside of the entire system. This approach makes it possible to capture and to balance all kinds of energetic processes and interacting forces.

As explained earlier, masses can be considered to consist of combined $1/c$-elements that have been produced by incredible strong forces and banned this way into the wrong space-time quadrant. Gravitation can be considered as a homing beacon force that allows these banned elements to return to their own space-time energy quadrant via enormous mass concentrations of black holes that are nothing else but tidal energy whirls and the doors back home.

VIII) References

[1] Special theory of relativity:
Albert Einstein, *Zur Elektrodynamik bewegter Körper, Annalen der Physik und Chemie.* 17, 1905, S. 891–921

[2] General theory of relativity:
Albert Einstein, *Annalen der Physik*, 49, 1916, S. 769-822

[3] Constancy of speed of light:
First experiment by Michelson-Morley in 1887
http://www.relativitycalculator.com/Albert_Michelson_Part_I.shtml

[4] A Treatise on Electricity and Magnetism Vol. 1 and 2
James Clerk Maxwell, 1904-edition with corrections – Antique Books Collection

[5] Raum und Zeit
Hermann Minkowski, lecture at the 80th Naturalist Assembly in Cologne, Germany on the 21st September 1908

[6] New significance for the cosmological constant of the general theory of relativity
http://hubblesite.org/newscenter/archive/releases/2009/08/

[7] Dark energy in astrophysics
The Cosmic Triangle: Revealing the State of the Universe
Neta A. Bahcall, Jeremiah P. Ostriker, Saul Perlmutter, Paul J. Steinhardt
Science 28 May 1999: Vol. 284. no. 5419, pp. 1481 - 1488 DOI: 10.1126/science.284.5419.1481

[8] Uncertainty principle
W. Heisenberg (1930), Physikalische Prinzipien der Quantentheorie (Leipzig: Hirzel). English translation: The Physical Principles of Quantum Theory (Chicago: University of Chicago Press, 1930).

[9] Title: Measurements of Omega and Lambda from 42 High-Redshift Supernovae.
Author(s): Perlmutter, S.
Source: The Astrophysical Journal 517 (2): 565-586
Published: June 1999

[10] Title: Dark Energy in Astrophysics. The Cosmic Triangle: Revealing the State of the Universe
Author(s): Neta A. Bahcall, Jeremiah P. Ostriker, Saul Perlmutter, Paul J. Steinhardt
Source: Science: Vol. 284. no. 5419, pp. 1481 - 1488 DOI: 10.1126/science.284.5419.1481
Published: May 1999

[11] Title: Observational Evidence from Supernovae for an Accelerating Universe and a Cosmological Constant
Author(s): Riess, A.
Source: The Astronomical Journal 116 (3): 1009-1038
Published: September 1998

[12] Title: New Significance for the Cosmological Constant of the General Theory of Relativity
Author(s): NASA, ESA, Riess A.
Source: http://hubblesite.org/newscenter/archive/releases/2009/08/
Published: May 2009

[13] Title: The Quantum Vacuum and the Cosmological Constant Problem
Author(s): Zinkernagel H.
Source: Studies in History and Philosophy of Modern Physics 33: 663-705
Published: 2001

[14] Title: General Relativity: An Introduction for Physicists
Author(s): Hobson MP, Efstathiou GP, Lasenby AN
Source: Cambridge University Press, page 187. ISBN 9780521829519
Published: 2006, 2007 edition with corrections

[15] Title: Geometric Representations of Lorentz Transformation
Author(s): Brehme RW.
Source: American Journal of Physics   Volume: 32   Issue: 3   Pages: 233-&
Published: 1964

[16] Title: A Geometric Representation of Lorentz Frames for Linearly Accelerated Motion
Author(s): Brehme RW.
Source: American Journal of Physics   Volume: 31   Issue: 7   Pages: 517-&
Published: 1963

[17] Title: A Geometric Representation of Galilean and Lorentz Transformations.
Author(s): Brehme RW.
Source: American Journal of Physics   Volume: 30   Pages: 489-496
Published: 1962

[18] Title: What is the universe made of?
Author(s): NASA, Hinshaw, Gary F.
Source: http://map.gsfc.nasa.gov/universe/uni_matter.html
Published: January 2010

[19] Title: Wilkinson Microwave Anisotropy Probe (WMAP) 7-year results
Author(s): NASA WMAP Team
Source: http://wmap.gsfc.nasa.gov
Published: October 2010

[20] Title: Zeitintervall in bewegten Systemen. English translation: Time Intervals in Moving Systems
Author(s): Stöcker H.
Source: Taschenbuch der Physik, Wissenschaftlicher Verlag Harri Deutsch, Frankfurt am Main. Page 137
Published: 2005

[21] Title: Broken Symmetries, Massless Particles and Gauge Fields
Author(s): Higgs P.W.
Source: Physical Review Letters 12, 132-3
Published: 1964

[22] Title: Broken Symmetries and the Masses of Gauge Bosons
Author(s): Higgs P.W.
Source: Physical Review Letters 13, 508
Published: 1964